Savage Factory, Cotton to Canvas, by Water & Steam

Patrick H. Stakem

(c) 2017

Table of Contents

Introduction..3
Author..4
The Savage Mill..5
Timeline..8
The Raw Material...17
The technology..19
The process..20
Mill Workers...24
The Market...25
The Clipper Ship..26
    The Baltimore Clipper......................................27
The Savage Rail Road...28
Competition..30
Compare & Contrast – Silk......................................30
Afterword..32
Bibliography...33
Glossary of Terms...39

Introduction

This book covers the topic of cotton production in Maryland in the 18th century Maryland, using Southern cotton. The process continued into the 20th century. The technology for automating the process was developed in England, and perfected in New England. Industrialization was not well developed in the South, but the region was good for cotton growth. Export of cotton to Northern factories, England, and France was a major money-maker for plantation owners. This book focuses on the Savage Factory I the Town of the same name, near Laurel, Maryland. It touches upon other facilitys, one in Cumberland, and several in Baltimore.

Southern cotton came by ship to the Northern ports, Baltimore for the Maryland region. It went to the Savage Mill via the Baltimore & Ohio Railroad's Washington Branch, completed in 1835, to *Savage Station,* just south of Annapolis Junction, and the current Route 32. From there, it went by wagon to the Mill, about a one mile journey. Before the railroad was complete, the cotton made the trip by wagon in both directions. Most finished product went to Baltimore. The Mill later had a warehouse to store the incoming raw cotton, and the finished canvas product. Improving "the final mile" was a priority for the mill, but a hard sell to the railroad.

Savage Mill is a facility along the North bank of the Little Patuxent River, in the Town of the same name. It was a working mill from 1822 to 1947. There was adequate water flow most of the year, and the area was used for water-powered mills since the middle of the 18th Century. Now, it is an upscale shopping and tourist center, in the midst of the remnants of 19th Century technology. It is adjacent to hiking trails that weave through the history of the area. The Town of Savage will a mill town back in the day, hosting a large manor house, and numerous examples of worker housing. It was an enclave of the Industrial Revolution in the United States, located in the middle of nowhere, with limited access. At the same time, it saw the war of 1812 fought to the south in Bladensburg and to the north in Baltimore. It saw the Civil War, which shut off its cotton supply in a time of increased demand, and World War I and II, which peaked demand for its product.

U. S. Route 1 and the Washington Branch of the Baltimore and Ohio Railroad, would be built near the mill. In the Industrial Revolution, transportation and sources of power were issues. Many facilities would be located near the source of raw material, but struggled with power sources until the perfection of the steam engine. Then, you had the issue of getting coal. The Savage Facility had all the free power it needed from the Patuxent River. It was not near the source of its raw material, cotton, nor very convenient to its major customer, the sail makers of Fells Point in Baltimore. We'll see how the mill operated, how the technology was updated, and how the transportation issued were addressed.

The milling process has its own unique vocabulary. A visit to the Mill Museum in Laurel, Maryland, will help with some of that, and this book has a glossary of milling and cotton terms. The numerous pictures and diagrams of the old milling machinery at the museum is also useful to see.

Author

The author resides in Laurel, Maryland, home to a water-powered cotton mill museum, and near the Savage Mill. He is fascinated by the Industrial Revolution in America, particularly the transition from wind and water to steam. His previous works have covered the Industrial Revolution in Western Maryland, particularly the Iron and Railroad manufacturing industries, and the Silk and Cotton Mill. He has degrees in Electrical Engineering, Computer Science, and Applied Physics, and no formal training in Mechanical Engineering or Cotton Production.

As part of the research for this book, he visited the ship and sail manufacturing facility at Fell's Point in Baltimore, and the Mills of the Jones Falls, as well as trips to the Laurel Museum and the Savage facility. He also managed to get some cotton seeds, and verified that cotton plants will not flourish north of the Potomac.

The Savage Mill

Alexander Warfield built a mill along the falls of the Patuxent near the Town of Savage, Maryland, in 1750. He left it to his sons. It was never much of a success, and was later sold to Francis Simpson.

The land was part of Anne Arundel County, and was acquired by Commodore Joshua Barney. The Commodore Joshua Barney House in Savage was built in 1760. Nathaniel Williams was a Commission Agent in Baltimore, who married Barney's daughter Caroline. He realized the potential of the area, and got his brothers Amos Adams, Cumberland Dugan, Benjamin, and George interested in building a mill. The War of 1812 interrupted the project, with brother Nathaniel wounded at the Battle of North Point, and brother Barney wounded at the Battle of Bladensburg.

There's an interesting tie-in between Amos and George Williams, and A. D. Hollingsworth of the Gunpowder Copper Works. George was a cotton and a copper agent. Levi Hollingsworth took over copper production just before the War of 1812. He also had an interest in several privateers, with government approval to fight and capture British ships. Another partner was Amos Williams, who had a vested interest in selling sails.

The Company was incorporated by an Act of the Maryland General Assembly at the December Session, 1821. Now, the Mill project needed an "angel investor" to put up the money for the project.

Benjamin had become wealthy in the shipping industry in Baltimore, but still couldn't finance the entire project. He reached out to his friend and relative John Savage of Philadelphia, who loaned them the money. The mill and surrounding town were named for him. Savage had been born in Kingston, Jamaica. He was successful in the shipping business, and formed the firm, Savage and Dugan in Philadelphia in 1791. He could see the potential of a mill that used product shipped in by sea, and produced sailcloth. Savage advanced the Williams brothers the money to start the business and later bought the mill outright in 1823 for $6,667.67.

The original facility included the mill, 500 acres of land, a warehouse, flour mill, saw mill, and later, an iron furnace and forge. The Town that built up around the Mill was mostly company housing, and it was named after the man who had bankrolled the project.

In 1824, a $12,000 loan from the Bank of Baltimore was used to expand the mill operations. The facility was already known as a premier manufacturing center, not just of cotton products, but also of mill machinery. An iron furnace on the site cost $8,000. It was not owned by the Company, but by Amos and Cumberland Williams, and Thomas Landsdale. By 1825, there was a warehouse, a flour mill, and a saw mill, presumably water powered. By 1825, men, women, and children tended 120 powered looms. Later, a grist mill was added as well as a iron foundry and a machine shop. In 1836, the company provided slabbing engines, turning lathes, and gear cutters to the Harper's Ferry Armory. Besides the Springfield Armory, this was the Country's main manufacturing site for muskets, rifles, and pistols. It was also powered by water. Some of the gun smithing machinery still exists and is on display at the National Park at Harper's Ferry. It is unknown whether any of this came from Savage. Most of Harpers Ferry manufacturing machinery was carried off and sent south by rebel troops in the Civil War. It first went to Richmond, then further south, and has been lost to time.

In 1829, Amos Williams acquired three adjacent properties in the area, and expanded the Mill complex to 980 acres, at a cost of $12,000. Amos was the Mill manager at the time, operating out of an office on Calvert Street in Baltimore. A dam was constructed on the Little Patuxent, less than a mile upstream from the Mill, below the confluence of the Middle Patuxent and Little Patuxent. A mill race was on the North side, where the mill was located. The dam had a water gate to allow for controlling the flow.

The main section of the dam in the middle was removed in 1950, but the southern and northern sections are intact. To reach the southern section, don't bother to walk upstream along the banks. It quickly becomes impassible. Take the Savage Mill Trail, across the Bollman Bridge from the Mill. It deteriorates, but continue until the dam section can be seen on the right. It is of masonry construction. Across the river can be seen the northern section. In the river can be seen some of the stones used to build the dam, much different than the native stone in the region.

The northern wall of the dam is reached by the Patuxent Branch trail, from Savage Park. The trail leads you to the water, and you hike downstream to the remains of the dam. The other section can be seen across the river, in the foliage. The north section leads to the mill race, and that continues down to the mill. The section close to the mill is occupied by an Adventure Sports area, with climbing and ziplines. At the upper end, the millrace is heavily overgrown.

In 1847, the entire Savage operation was sold to William H. Baldwin of Woodward, Baldwin & Co. of Baltimore and New York. Co-purchasers were Baldwins's brother Christopher and H. N. Gambrill. William built a house in Savage for his family, in close proximity to the mill. His four daughters worked at the mill. His retirement home is on the Maryland Historical Trust's Maryland Inventory of Historic Properties, but was scheduled for demolition in 2013. It was located at 8334 Savage-Guilford Road.

The oldest remaining mill structure is the stone carding and spinning building, probably built between 1816 and 1823. The mill was expanded by the Baldwins before 1881, and that expansion included the brick tower in Romanesque style. Other buildings include the weaving shed, cotton preparation area, paymaster's office, and several early-20th century warehouses and power plants.

Timeline

1750   Alexander Warfield builds a mill along the falls of the Patuxent.

1760   Commodore Barney House built in area that will become Savage.

1779   Spinning Mule invented in England.

1791   Savage & Dugan Company formed in Philadelphia.

1810   Eleven cotton mills in Maryland

1811   Nicholas Snowden begins mill operations in Laurel

1812   War with Britain; Battles of Bladensburg and North Point in Maryland.

1816   Carding and Spinning building built.

1820's Savage Manufacturing Co. formed, work on mill starts.

1820 Maryland produces almost 850,000 pounds of spun cotton from 44 mills

1821 Company chartered in Md.

1822 Mill in operation.

1823 John Savage buys the mill business from the Williams Brothers

1824 Mill operations expanded with a loan from the Bank of Baltimore.

1825 Warehouse, flour mill and saw mill added.

1827 Facility produces the Knowles wood working planes.

1829 Amos Williams is the Mill Manager, working out of Baltimore. Dam constructed.

1835 B&O Railroad Washington Branch opens; Amos & Cumberland Williams incorporate the Savage Railroad Company.

1836 Facility provides machinery to the Harper's Ferry Arsenal.

1839 Savage provides milling machinery to Whitehall Mill in Baltimore.

1839 Amos gives up management role to his brother, Cumberland.

1840 Mill facility produces "Baldwin's Cotton, Hay, and Tobacco" presses. America exports the most cotton.

1843 China trade increases demand for fast sailing ships

1846 Company involved in a legal dispute about milling machinery supplied to Powhatan Manufacturing Co. and claimed by the Bank of Baltimore over a loan issue.

1847   Operations and property sold to William H. Baldwin.

1850   Cotton picking machine invented. Previously, cotton was picked by hand

1859   Will Henry Baldwin buys the facility and takes over operations.

1862   Mill shuts down due to lack of access to southern cotton.

1870   Bollman Bridge installed at the mill.

1880   Mill makes the transition from water power to steam engine. Water Turbine provides electricity for the mill and the Town.

1881   Tower and other expansions. Pay master's office built.

1911   Company files a patent for drying paper using cotton duck.

1918   Company merged with Baldwin, Leslie & Co.

1942   World War-2 needs drive production to 4,000,000 pounds of cotton duck per month, 372 employed.

1947/48 Facility closes; becomes a Christmas ornament factory.

1950   Patuxent dam (center section) removed.

1974   Property placed on the National Register of Historical Places.

1985   Historic Savage Mill Partnership takes over. Facility opens as a home for boutique stores and restaurants. Most of the interior is original.

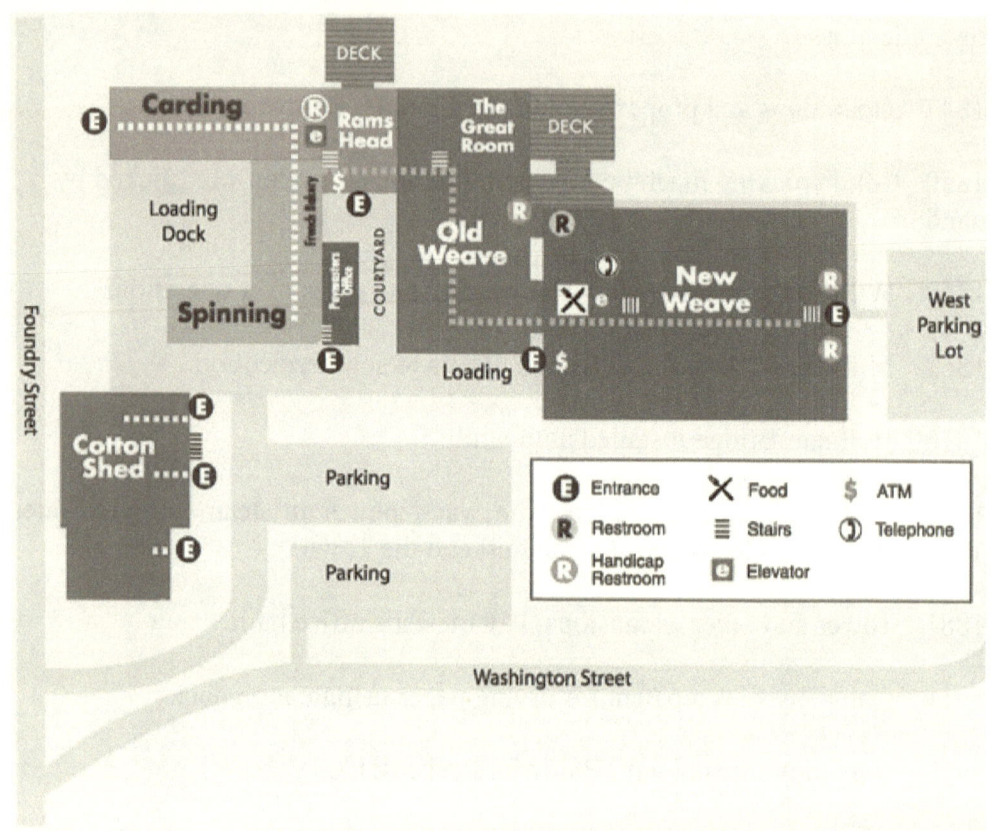

diagram, courtesy Historic Savage Mill, A. J. Properties.

To get you oriented, the Bollman Bridge is up from the carding building. The Pax flows from right to left, above (South) of the facility. The deck out from the Ram's Head Tavern gives you a good view of the steam power plant, out near the river. The deck from the Great Room give you a good view of the water power turbine facility.

The stone structure just past the Bollman Bridge is the original carding and spinning building, circa 1816 The four story brick Romanesque tower was constructed before 1881.

A popular attraction is the Adventure Sports area, featuring climbing walls and zip line, along the old mill race. That is to the right, in the diagram.

This was by no means the only cotton mill in the vicinity. Laurel Mill, just a few miles south, used the water power of another branch of the Patuxent. It was built by Nicholas Snowden on the site of an earlier grist mill.

From 1835 to 1851, it was operated by Horace Capron, who married a Snowden. He formed the Patuxent Manufacturing Company, and later added the Avondale Mill, in Laurel. Transportation was easier, as the B&O Railroad's line was at the other end of Main Street. The Laurel operations employed around 800. Capron also owned mills in New York.

The mill was powered, like Savage Mill, by a branch of the Patuxtent River. A town called Laurel Factory grew up around the mill, and was later incorporated as the Town of Laurel. The facility was demolished in the 1940's. It was located where the Town's municipal pool is now located. Walk upstream from the pool, and see part of the dam that was used to provide proper water flow to the mill. The mill operated until 1929.

Horace Capon married into the Snowden family, had operated Savage Mill. He formed the Patuxent Manufacturing Company, and operated Laurel's Avondale Mill, built in 1845 near the original Laurel Mill. The mill had 1,500 spindles in work, tended by 150 employees. Later, it was repurposed into a gristmill. The water wheel provided around 70 horsepower, and was later replaced by a steam engine, using coal from Western Maryland, transported by the B&O Railroad. The Avondale Mill burned in 1991, and the site is now used as a community park.

There were numerous Snowden Mansions in the area (most survive), on the Snowden's 9,000 acre plantation. Some members of the family operated an iron making facility, the Patuxent Iron Works, that exported product to England.

He served as the third Commissioner of the Department of Agriculture under President Jackson. He started his career in the cotton manufacturing business as a superintendent of a cotton factory in Baltimore. The Laurel Mill was operated by a water wheel, later replaced by a turbine. A small steam engine was added later. The factory made cotton duck and other products. A Mill/Textile Museum, and part of the dam, can be visited today at the very west end of Main Street in Laurel.

As we shall see, the Savage operation was not just a cotton mill sited near abundant water power. The facility had to produce it's own parts and machinery, there not being a "cotton mill kit," so the facility included the iron furnace, and shops for producing parts and machinery. The shops also produced machinery for other mills, and the U. S. Government's Harper's Ferry Arsenal. Being rather remote, the facility had to produce and repair whatever it needed from local raw materials.

In the Maryland Industrial Census of 1810, there were eleven cotton mills in Maryland, most in Baltimore, along the Jones Falls. By 1820, the 44 Maryland mills reported 50% utilization of their mills and spindles. In 1840, there were 50 textile manufacturers in the State, 19 of these in Baltimore, employing 2,672.

In 1831, Mr. Hack, a machinist at the Savage Mill, developed a machine for reeling and twisting silk that was submitted for a patent.

Mr. Horatio Gambrill was a weaving apprentice at the mill around 1828. Gambrill's wages were $104 for the first year, $154 for the second, and $204 in the third. He would eventually became rich in the textile industry, but not at those wages. In 1839, he converted the Whitehall Flour Mill in Baltimore to a textile mill, producing cotton duck. He had five looms for weaving ship's sails. He introduced Baltimore to cheaper sail cloth, without sacrificing quality. This was not seen as a competition by Savage Mill. In fact, they manufactured the mill machinery for him, and delivered it on long term credit terms.

The Savage Company offices were moved from Baltimore to the mill site in 1832. The Company address is listed as 411 Telegraph Road, in Odenton. By then, the facility also had a grist mill, saw mill, machine shop, foundry, blacksmith shop, wheelwright facility, brick-making capability, a company store, and the iron furnace. It was self-sufficient for mill construction and maintenance. Parts from the facility were purchased and used by Joseph

Bancroft to construct a cotton mill in Rockdale, Pa. That Town is situated along the banks of Chester Creek, in southeastern Pennsylvania.

The mill, like many early industrial sites, was constrained by lack of transportation, both for incoming raw material, and for shipment of finished product. The mill was sited where the water power to operate it was available. Water power is essentially free, once you build the infrastructure of the water wheel and dam. Of course, the water flow was not consistent. There were floods that damaged the machinery, and droughts that shut the mill down. This situation lead later to a steam option for power.

In 1839, Amos Williams gave up management of the company to his brother Cumberland, due to illness. Later, in 1848, Amos entered claims against the Savage Manufacturing Company and his brothers George, Cumberland, and Nathaniel. He claimed they tried to force him into poverty from indebtedness during his illness.

In 1843, additional land was purchased by the mill for firewood and ore for the furnace operations. The furnaces were on the North side of the river (on the the mill side and upstream). By 1835, there were two furnaces, with an even older iron foundry. They weren't used after 1839. There was also a cupola furnace, which was apparently never used.

In the early 1840's the facility made a machine, "Baldwin's Cotton, Hay, and Tobacco Press," which found an eager market in the South. This may have been the subject of Patents issued to Joseph C. Baldwin of Virginia. The facility also produced the patented Rechm planing machine.

The company produced wood-working bench planes, patented by Hazard Knowles of Connecticut, (number 4859x), dating from August 24, 1827. This is the earliest known patent for a wood plane. It came in a variety of sizes, and looks remarkably modern. It is made with a cast iron frame, and a wooden knob. Known examples approach $1,000 currently for tool collectors. The plane defined a series of later wood-working tools, "of the Knowles type."

It is not known if the Savage Manufacturing Company made other hand tools as well. There is currently a Savage Tool Collection from the Swanson Tool

Company of Frankfort, Il, but its relationship to the Savage Company is unknown.

In 1859, William Henry Baldwin Jr. took over operations as the Savage Manufacturing Company, purchasing the land and factory facilities for $42,000.

During the early part of the Civil War, the facility shut down, due to the lack of cotton from the Southern States. Cotton is grown south of the Potomac river, from Florida to Texas. Maryland produced a small amount on the Eastern Shore, not nearly enough to meets the needs of industry, let alone a wartime surge in demand. The mill shut down until cotton could again be obtained.

By 1880, the facility transitioned from water power to steam, for more consistent operations. It was no longer dependent on the fickleness of the Little Patuxent. This was not uncommon among the country's mills and manufacturing facilities of the time, particularily in New England. It is quite possible the steam engine was built on-site. Some early cotton mills were powered by the wind, very Green, but not too reliable. The flow of the old mill race was used to power a long drum arrangement, which drove a generator that provided electricity to the mill and the town. Electric lights were a productivity enhancer.

From 1905 to the end of World War-I, the mill was managed by William's son Carroll. In 1911, the Company filed a patent for a Process for drying paper using its cotton Duck product. Sales of sails for Clipper ship were down in the age of steam. In 1910, the facility was owned by Woodward, Baldwin & Co. Baltimore.

The Company then merged with the New York-based Baldwin, Leslie and Company. In 1918 the company was renamed Leslie, Evans and Company after Baldwin's death. Harry M. Leslie managed the Savage Mill, Clinchfield Mill in North Carolina, a mill in Texas, and the Clinchfield Manufacturing Company in North Carolina. Woodward, Baldwin & Co. was a textile manufacturing company, that benefited from having its own cotton mill. They made bed sheets, pillow cases, and piece goods from cotton, and, later, rayon.

They also produced muslin, and drapery liners. They had a series of trademarks for their products.

Interesting information about the company and its operations can be found in the Maryland State archives. In, *Reports of Cases in the High Court of Chancery of Maryland 1846-1854*, there is a case, "Williams vs. Savage Manufacturing Co." This relates to disagreement among the Williams Brothers. Williams, the plaintiff, claims he was not paid the correct amount of dividends for his stock in the Savage Rail Road Company. There is also a dispute of who came up with the funds to erect the iron furnace, which is listed at $8,000. There was an agreement, dated June 1, 1844, which seems to at least partially settle the issues, but it is reported that it was not signed by the requisite number of shareholders. Makes for an awkward Thanksgiving at the Williams'.

The most interesting source of information is from *Reports of Cases in the High Court of Chancery of Maryland 1846-1854*. There is a lengthy discussion about the Powhatan Manufacturing Company and its dispute with the Bank of Baltimore, about a construction loan. The Bank was foreclosing on the Powhatten building, some 6 miles from Baltimore. The issue was the machinery, which had been installed. The argument was whether the machinery was part of the building. This included a steam engine, boilers, tanks, and appurtenances from the Savage Manufacturing Company, that hadn't been paid for.

There is also a relevant section in the *Court of Appeals of Maryland*, regarding Savage Manufacturing Company v. Owings, 3 Gill 497 (1846). It's unclear what the issue was, but it involved an act of the General Assembly to open a new public road.

Canvas production was strong at Savage through the end of World War-1. Lack of demand after that, plus the demise of the age of sail, led cotton producers to look at other markets, such as wicks for oil lamps, sash cord, and netting. The railroad discontinued service on the Patuxent rail spur in 1928, but the Savage Mill could still be served. Times were tough during the Great Depression, for cotton, and all over. Demand was dropping for utility cotton products. There was a brief upturn in the market during World War-II. The mill closed in 1948, and became a Christmas ornament factory.

In 1975, an attempt was made to reclassify the mill from industrial to business. A restoration took place up until 1985. The rezoning effort was not successful. It did get listed on the National Registry of Historic Places, which provides protection from destruction.

In 1985, Savage Mill Limited Partnership took over the site, and opened the mill facility for restaurants, boutique shops, and antique dealers. The partnership had procured State and County money for the Project. The site had been fairly successful, with visitors reaching one million by 2010.

We are truly fortunate that the Mill facility has survived. It is now a complex of 12 buildings, with 175,000 square feet of space. Interpretive signs give a hint of the buildings' prior uses, allowing your imagination to hear the rumbling of the machinery, as raw cotton came in one end of the facility, and woven cloth went out the other.

The Raw Material

Cotton has been in use since Neolithic times, some 8000 years. It is a natural vegetable fiber that grows in a protective boll around it's seeds. It is made up almost entirely of cellulose, which is a structural material of cell walls in plants. The cotton plant can be found in tropical and subtropical locations anywhere in the world. More than 100 million bales are produced every year, and it is the most widely used natural fiber.

Cotton was hand-picked up until 1850, when a machine was developed to carefully pick the boll without shredding the fibers. The invention of the cotton gin to separate the seeds from the fiber was the key invention needed to allow the wide-spread industrial use of cotton for cloth. Manually, it took about 600 hours to separate out the seed from a bale of cotton. Cotton is still picked manually in some developing regions.

Cotton seeds are not a waste product, but are used to make cottonseed oil. It can be used like any vegetable oil, even for human consumption.

There are many species of cotton, but most commercially grown product is one of four types. These are native to the southern United States and Central America, South America, India and Pakistan, Southern Africa and the Arabian Peninsula. U.S. Cotton now represents 90% of world production.

Herodotus wrote that "..in India, trees grew wool..." In the middle ages, there was speculation that it was the product of plant-borne sheep. The British East India Company brought a huge trade in cotton from Asia to England. America took over the lead in cotton exports by the 1840's.

Before the Civil War, a former mill manager saw which way the winds were blowing, and bought as much cotton from the South as he could. He supposedly made a profit of $80,000 when the cotton supply was cut off by the war.

Generally, cotton doesn't grow above the Potomac River. I confirmed this, or maybe it was my lack of a green thumb. Some grew East of the Chesapeake Bay, but that was, for all intents and purposes, Southern territory. Little or no cotton was processed in the non-industrialized South, but they sold their raw material to England and France to provide revenue for the Confederacy. An active blockade was set up by the Navy, but cotton still provided a source of hard currency for the government in Richmond.

At the same time the supply of raw material was cut off, the demand for cotton products for the Army and Navy soared. This included tents, sails, wagon covers, etc.

The technology

The Savage Mill was initially powered by a thirty foot diameter water wheel, operating from a constructed millrace about a mile up along the north side of the river. A dam was constructed to divert water into the mill race. It was removed in the 1950's. The power produced by the wheel depends on the flow of water, and the drop, or head. The wheel is not 100% efficient, but usually about 70%. It converts the water flow (kinetic energy) and water drop (potential energy) to useful work. A simple equation gives you the horsepower of the wheel. The flow varies throughout the year. This also assumes the water is introduced at the top of the wheel. Looking at the U. S. Geological Survey data for the Little Patuxent at Savage, for March of 2017, shows a flow of between 60 and 140 cubic feet per second. With a 30 foot wheel, they could get around 170 horse power to operate the mill machinery. The mill shaft went through the "Old Weave" building and drove the looms and spinners, and in the carding building, with a system of lineshafts at the ceiling, leather belts, and pulleys. When the New Weave building was built in 1916, the mill race was directed to a flume, at the current location of the deck at the end of the *Great Room*. It passed thru a wall of iron bars acting as debris screens, and drove a drum, 12 feet in diameter and 30 feet long, constructed of cast iron, possibly on-site. The water-powered drum drove a electric turbine that powered the mill and the entire town.

When the mill was converted to steam power, it necessitated getting access to coal. The line-shafting and belts to operate the machines could be left intact. This time, water at a higher temperature drove the shafts. The steam engine is not as efficient as the water wheel, but is more consistent in its power delivery. The brick building that is adjacent to the deck of the Rams Head Tavern and out in the water, housed the steam engine.

The spinning mule, so called because it was a hybrid of Arkwright's water frame and Hargreaves' spinning Jenny, was invented in England in 1779. A water frame is a spinning frame, powered by water.

With powered machinery, mill workers became less of a skilled craft than machine attendants. Spindles were used in conjunction with a frame, such as the water frame of Arkwright, discussed later.

The process

Cotton bolls, or seedpods, were removed manually from the plants. Each seed in the pod, and there can be thousands of them, has about an inch of fiber attached. The seeds have to be removed. Originally, this was also done by hand, a tedious process. After the invention of the cotton gin (engine), this process was mechanized. The cotton "lint" is then compressed into bales, of typically 500 pounds weight. These were delivered by wagon, later by train, to the cotton mill.

The raw cotton arrives with residual seeds, and vegetable matter. Some of this can be picked out by hand, a job that was usually reserved for children. A powered *Scrutching* machine, invented in 1797, passes the cotton through a pair of rollers, and then strikes it with beater bars. This process knocks most of the residual seeds out. Then it's on to the carding machine.

The carding machine has multiple rollers, where the fibers are aligned to make them easier to spin. The machine has one big roller with smaller ones around it. All the rollers have small protrusions, called *teeth*. As the cotton proceeds through the machine, the teeth on the rollers get closer together. The cotton emerges as a sliver, a rope shaped fiber. The stone carding and spinning building at Savage Mill was the center of all activity, until the mill was expanded in 1881. It was originally 3 stories, later expanded with a fourth. Each floor was one large room. The building had a stone tower with stairs, topped with a bell tower. The bell in the Tower would be used to signal a fire or other emergency.

An expansion building, the weaving shed was to the west, upriver, from the spinning and carding building. It is two stories, and has light wells on its roof. The big warehouse was added in 1916.

Drawing is the process of producing uniform strands of fibers by passing them through another series of rollers. The strands can be made as thick as desired. Slivers can be combined into a more consistent *rope*. The slivers are separated into *rovings*. The rovings head to the spinning machine. Spinning refers to twisting multiple fibers to make yarn. The fibers are wound on a bobbin, which is held by a spindle. Multiple spindles reside in a frame, and they are replaced as they fill up – a manual, repetitive, simple task.

Savage Mill had both Mule spindles, and Ring spindles. Essentially, the spindle machines take the rovings, and spins them, creating a yarn. The mule spindle generally had a "minder" with two boys to help. It was developed in 1790, and is still in limited use for exotic fibers such as cashmere, merino, and alpaca.

A mule spinner takes the roving from a bobbin, and feeds it through rollers that are operating at different spin rates. This step thins the roving to a consistent size, and makes a cone-shaped bundle of fibers. Mule spinning produces finer thread than ring spinning, and a less skilled operator can be used for mule spinning. The spinning mule was the work of William Compton of England, who combined the previous Water Frame of Richard Arkwright (1769) with the rollers of the Spinning Jenny of James Hargreaves. Arkwright developed in England a spinning frame powered by water. Semi-skilled labor could now be used. He is credited with being "the Father of the modern industrial factory system." His contributions to the Industrial Revolution in England were huge. He had been apprenticed to a barber, and developed a waterproof dye for the wigs then in fashion. This would fund his later inventions. He worked on spinning and carding machinery with John Kay, a clock maker. Their machine substituted wooden or metal cylinders, turned by machinery, for worker's fingers. He set up water powered mills filled with his inventions, and producing cotton faster and cheaper than others. His spinning frame was a significant improvement over Hargreaves' earlier spinning jenny. It was easy to master, and required less skilled operators. Samuel Compton would later combine the water frame and the spinning frame into the spinning mule.

The capacity of a mill depends on the number of frames, and the number of spindles in frames. Savage mill is said to have had 12,000 ring spindles, 5,000 twister spindles, and 72 cards at its peak. All of these were turned by leather belts from overhead line-shafts, powered by the water mill. The machines were serviced by a series of semi-skilled workers, usually women and children. The industrial workplace was extremely dangerous. Lives and limbs could be lost in seconds. The idea of Industrial safety was unknown. Workers were easy to come by. The machinery was noisy, and the air was filled with cotton fiber dust. The only light came from the windows. Fire was always a hazard, and in certain concentrations, cotton dust is explosive, so open flames were not used. Before electricity, your choice was daylight.

A loom is where the yarn from the spinners is made into cloth. Savage Mill had 194 powered looms, turning out heavy cotton duck, or canvas. The threads running the length of the fabric are called the *warp*, and cross threads are the *weft*. The weft passes across the loom on a shuttle carried on a pirm, This is a weaver's bobbin, or spool.

Cotton thread from the spinners is taken to the warping room, where the correct length is wound onto warper's bobbins. On the loom, there are racks of bobbins. Sometimes, the warp is *sized*, or treated with a starch to cut down on breakage.

In the time-frame the mill was active, one person could manage 4 looms. All of the machines could individually be disconnected from the power supplied by the overhead line-shaft, by lifting a pulley to loosen the drive belt. When the looms are operating properly, the attendant just watches them work. When something goes wrong, like a broken warp thread, a broken weft thread, or a machine jam, action is needed in a hurry. The first step is to disconnect power to the machine, to prevent further damage. Then the problem can be resolved, the mess cleaned up, and the machine put back into production.

On the loom, the warp is divided into two lines, so the shuttle can pass between them. Picking refers to the process of projecting the shuttle from one side of the loom to the other, through the warp threads. Then third step is "beating up," where the reed, or comb, separates the warp threads, and pushes the weft yarn into place.

The Jacquard loom, none of which are known to have been used at the Savage facility, allowed an automated process, using punched paper cards, to control a sequence to weave intricate patterns. Once, a portrait of the inventor was done in this process. The cards later influenced the development of punched cards for early computers.

Picks refer to the weft and *ends* refer to the warp. The coarseness or fineness of the cloth can be expressed as the number of picks and ends per square inch. The canvas produced by Savage Mill was a course material, and was probably not bleached, treated, or dyed. It would typically weigh 30 ounces per square yard.

The mill in 1941 had 194 power looms, 12,000 ring spindles, 5,000 twister spindles, and 72 cards. One special loom could make canvas 208 inches wide. The sheer number of looms was useful for World War-II production demands. The Mill's output increased to 400,000 pounds a month.

Throughout World War II the mill produced heavy duck for canvas, hose, sails, and community electrical power. During the War, the Mill was putting out 4,000,000 pounds of canvas per month. The demand for canvas dropped considerably after the war, and the mill was scheduled to be shut down on 1 January 1948. At the time, the 400 acre complex employed 372, consisted of twelve industrial buildings and 98 houses owned and rented by the mill to the workers.

Mill Workers

In the early Industrial Revolution, working conditions were not very good. That's putting it mildly. Mill owners knew there was a large available labor pool. Things were the same at the iron furnaces and coal mines.

At the mill, most jobs required agility, skill, and not much heavy lifting, so many women were employed. Children also were used for menial tasks. Skilled labor was needed for the machinery, and strong workers to unload the cotton bales, and load the canvas. Workers from the factory worked 6 days a week in 10 hours shifts.

Workers in the factory were on lock-down throughout their shifts, posing a risk if a fire broke out. Cotton lint in the air is flammable, and the workers breathed it all the time.

The lock-down resulted in a problem, because the workers couldn't go outdoors to "use the facilities." This was solved by adding bathrooms on each floor. May I suggest you dine on the deck of the Ram's Head Tavern. As you look back at the building, the bathrooms are to the right of the door you came out. Notice they are on each floor. They have been re-purposed as ... bathrooms.

A messenger would deliver notes if communications with those inside was needed or they could speak through a small window in the locked door.

The mill buildings had very large windows, because there was no interior lighting. Electricity for lighting was not used in the early days of the Mill, and oil or gas lamps would be a fire or explosion risk. On rainy or overcast days, you just had to make do. Working around all of that moving machinery was a major risk.

As with most early manufacturing centers, the employees were paid in script, acceptable only at the company store, If you were fired from your job, your script was useless. You also had to vacate company housing. This was the period of Industrial Feudalism, where workers were "vassals" of management. Savage was a company town, and the Savage Manufacturing Company was nowhere near the worst in treating its employees.

The Market

Most of the canvas from Savage Mills went to Fells Point in Baltimore, to be made into sails for the Clipper ships. Fell's Point is Baltimore's deep water port, and hosted numerous shipyards and supporting industries. Two of the first ships of the U. S Navy were built there, the USS Constellation and Enterprise. It was also an emigration port (probably for the author's ancestors), second only to Ellis Island. The area is on the National Register of Historic Places. In 1809, it had nine shipyards and 11 sail makers. In 1843, demand for ships was driven by the emerging China tea and opium trade.

Another market was the Army, who needed tents, wagon tops, and cannon covers. Canvas was also used as backing for paintings, and for large painted backdrops for stage plays. Later, canvas sea bags accompanied U.S. Sailors around the world during two World Wars.

Sails are made from the heavy cotton duck, a corruption of the Dutch word *Doek*, which means cloth. Baltimore, home of the famed Clipper Ships was a huge market for sails. The more sails you could cram on a ship, the faster it would go, and the quicker it would reach its markets. Nice thing for the sail makers and mills was that sails wore out and needed to be repaired or replaced.

Cotton, unfortunately, is not ideal for the maritime environment. It is subject to ultraviolet light degradation, water absorption, and resulting rot. Sails came to be made of cotton in the U.S. as flax had to be imported, but cotton was native. Linen sails, were preferred, but became too heavy when wet. Carded cotton sailcloth was traditionally 23 inches wide. The quality of the weave is important, as the correct weave controls stretch.

A sail plan would be drawn up for each ship by a Naval Architect, and defined the various combinations of sail for the ship. Different sail types are used for sailing into the wind (tacking), sailing before the winds, for heavy versus light winds, and storm sails. Spare sails and spare canvas was also carried. The sail plan for the clipper *Pride of Baltimore-II*, which can seen, toured, and sailed upon from Baltimore's Inner Harbor or Fell's Point, defines 9,327 square feet of sail.

A Sail-maker is a skilled trade for making and repairing sails. Although ships would carry someone who could repair sails, construction and major repairs were done in a sail loft facility on shore.

A Sail-maker's needle is similar to leather wedge needles, but the triangular point extends further up the shaft. It is designed for sewing thick canvas.

The Clipper Ship

The Clipper ship developed in the early to middle 19th Century as a fast trading & cargo vessel. Time was money, and the ability to reach markets ahead of the competition was essential. They were designed narrower than other sailing ships of the time, and could carry more sail. They sailed anywhere in the world, in specific trade routes. The apex of the Clipper was 1843, and the tea trade with China. Shortly after that, gold was discovered in California, increasing the demand to get to California from the East Coast.

In archaic use, *clip* means to run or fly swiftly. *Clip* began to be used to designate speed for ships and horses. The usage may have originated in Baltimore. The Clippers were optimized for speed, and suffered in cargo capacity. They carried many extra sails, and required additional crew to manage them. They operated at the limits, sometimes with the top of the lee rail under the water. They also had a relatively short working life of 2 decades or less. They generally weighed 200 tons maximum.

Because of their limited cargo space, the Clippers mostly addressed high value cargo such as spices and opium. They also had a role in the war of 1812, where the swift armed Clipper Ships thwarted the British to attempt to invade Baltimore, after burning Washington. That did not work out well for them, but we got a National Anthem.

Alan Villiers says, "To sailors, three things made a ship a clipper. She must be sharp-lined, built for speed. She must be tall-sparred and carry the utmost spread of canvas. And she must use that sail, day and night, fair weather and foul "

The Baltimore Clipper

The Baltimore Clippers were topsail schooners, developed before the American Revolution for the Chesapeake Bay. They were fore-and-aft rigged, with two or three masts. They are gaff-rigged, meaning they use triangular sails, missing their top point, The upper edge of the sail is fastened to a spar called a gaff, When close-hauled to the wind, the top of a sail tends to twist away from the wind. The gaff prevents this, as well as reducing the angle of heel the ship presents. (How far it leans). Fore-and-aft rigging is easier to manage than other forms, when the ship changes course. Multiple jib sails are used in front of the foremost mast. These are triangular sails that act as an airfoil, increasing the ship's performance, and reducing turbulence on the leeward (downwind) side of the main sail.

Clipper ships were used as warships in the Revolution against Britain, and in the War of 1812. They were small and maneuverable, usually mounting 12 guns. They could easily outmaneuver larger British war ships in the Chesapeake Bay. They were also used as commerce raiders off the British Isles, operating under Letters of Marque and Reprisal.

An example, the *Chasseur*, built at the Kennard & Williamson shipyard at Fells Point in Baltimore, was used to run the British blockade of the Port. The *Ann McKim* may have been the first of the Baltimore Clipper Design in 1833. She influenced the design of the extreme clipper, the *Rainbow*. This ship was built to the design of American Naval architect John W. Griffiths. He is considered to be "the most remarkable of America's 19th Century shipbuilders." His masterpiece is considered to be the *Sea Witch* of 1846. She set a record of Hong Kong to New York, around South America, in 74 days,

14 hours in 1849. As of 2013, that record for single hulled craft still stood. Griffiths later published "Treatise on Marine and Naval Architecture" in 1850. He never made the transition to iron ships and steam.

The Savage Rail Road

The Mill always had a transportation problem. It was somewhat off the beaten path, a location chosen for its access to water power. The raw cotton came in by train from the south, and the finished canvas went on the B&O Railroad north to Baltimore. The Washington Branch of the B&O line ran over on the east side of current Route 1, with a station and connection with the Annapolis and Elk Ridge Railroad at Annapolis Junction. After failing to convince the B&O to build a spur to their factory, the owners of the Mill decided to charter a railroad of their own in Maryland.

In 1835, two of the brothers (Amos and Cumberland) incorporated the Savage Railroad Company to allow connection with the B&O Railroad, replacing wagons on the bad roads. The spur line to the B&O connected south of Annapolis Junction, and just south of the Patuxent Bridge.

The B&O was not going to spend money putting in a line to a single customer. But they were willing to support the Savage RR Co. to do it. That line (now removed) branched off the B&O Main just south of the B&O's Patuxent River Bridge, called Savage Station. The mainline bridge is still in use, but the Savage spur bridge has been removed. It is unclear whether there was an actual station or platform at the location. Remnants or the Savage RR in terms of rail embedded in blacktop and piles of ties can be found at the back of the Savage II & III Industrial Park, off of U. S. Route 1. Further north on Route 1, the Poist Gas Company used the rail spur in recent memory for incoming delivery of propane tankers

In the 1870's an iconic Bollman Truss Bridge was used to cross the Little Patuxent to get to the warehouse facility. This bridge, the invention of Wendel Bollman of the B&O, was a game changer. It was patented in 1852. Previously, when a railroad bridge needed to be built, a work gang would go to the site, cut down trees, and build a bridge. Bollman's genius was that his bridge parts were built of iron at the B&O Factory in Baltimore, and transported to the site on railcars. Then the work gang would prepare stone footings, and assemble the bridge on site. The particular one at Savage Mill had been in use at another location, and was disassembled and shipped to Savage for re-assembly. It has its own space on the National Registry of Historic Places, and is the only known surviving example of its type. The bridge is decorated with lights in the Christmas season.

As you cross the bridge, you'll notice the cast iron compression members on the bridge pilings, and the wrought iron tension members in the overhead structure. Each of the tension members has a turnbuckle. As the bridge settles and ages, the turnbuckles are adjusted. A good bridge tuner could smack the tension member with a wrench, listen to the tone, and loosen or tighten that member properly. There are very few good Bollman Bridge tuners left.

The track extended to the warehouse building at the mill after crossing the river. The canvas cloth would be brought to the train from the warehouse, for backhaul to Baltimore. There was a restriction that a locomotive could not cross the bridge, probably due to weight. Pusher cars, empty flat cars, were used to move the loaded cars across the bridge. There is some evidence of a winch on site that would take the cars to their final destination. After the mill switched from water power to steam, the rail line was also used for coal delivery to the new power plant on the north side of the river. The rail line was used until the mill closed in 1947. The B&O had extended the line 2.5 miles to Guilford to service the granite quarries. That section of the line was closed in 1928, and is now a hiking/biking trail.

Before the B&O had built the extension to Guilford, granite traveled in wagons drawn by 6-horse teams, down a macadamized road to Savage Factory to load on the train.

Competition

At the Jones Falls in Baltimore, there were numerous cotton mills. There was plentiful water for all, and the nice thing is, it could be reused 1multiple times as it flowed down toward the harbor. Most of the mills were steam powered. Mill machinery was also built in the area. The Jones Falls goes from Rockland, north of Baltimore, to the harbor, dropping some 260 feet. There was a two mile segment where the current was ideal for power use. Major industries also included flour milling.

The first mill on the Falls was the Mount Washington Mill in 1810. This filled a need for home-produced cotton product, as President Jefferson had slapped high tarriffs on the imported British product. The mill had 1,000 spindles, 7 looms, and also a dying facility. The Whitehill Cotton Factory followed in 1839. They had five looms, producing duck for sails. The facility expanded with the construction of the Woodberry factory.

Other mills included Rockland, Park, Druid, Hoooperwood, Meadow, Clipper, and Mount Vernon. The Poole & Hunt Foundry provided machinery. Besides sailcloth, the mills produced military uniforms, tents, and wagon covers. By the late 1890's the Jones Falls Region had 70 percent of the U.S. market. This dropped to 60% before 1910. After World War-I, demand dropped, prices fell, strikes were called. Operations moved to the South, the source of the raw materials, with cheaper labor.

It was bad along the Falls, except for Hooperwood – he made the tents for the Barnum and Bailey Circus. He avoided labor conflicts.

Demand for cotton duck skyrocketed in World War-II, but again dropped like a rock at the end of the war. Cotton companies went out of business in the 1970's and 80's.

There facilities, much like Savage Mills, house restaurants, artists lofts, and boutique shops today. The Mill buildings can be toured today. (http://baltimoreindustrytours.com) Up to 80% of the cotton duck manufactured in the United States in the late 1800's came from along the Jones Falls. Yet the Savage Mill did well with its portion of the remaining 20%.

There were also many textile mills along the Jones Falls in Baltimore and other nearby locations, making good use of that source of water power. In the Census of 1820, the State of Maryland is listed as producing 849,000 pounds of spun cotton, using more than 20,000 spindles.

Compare & Contrast – Silk

This section will take a look at a silk mill in Lonaconing in Western Maryland, to explore the similarities and differences between that and the material and processes for cotton. The first thing to note is that silk is an animal product, and cotton is a vegetable product. The raw cotton had to be spun into long threads before it can be woven. The silk comes as a single long strand, up to 5,000 feet from a single cocoon. Once we have bobbins of thread, the weaving process is similar. Silk has been in use since 3,600 B.C. Silk has been found in the hair of a mummy in Egypt, dated to 1000 BC.

In 1907 George Klots of New York opened a silk mill in Lonaconing to take advantage of the inexpensive labor and cheap coal available in this coal mining area. The operation depended heavily on the labor of women and young children, who were not employed by the mining industry. Labor conditions improved through the years, and the mill continued production until 1957, when synthetic fibers encroached on the traditional silk fabric market.

Beginning with a crew of mostly youngsters, some as young as seven, the mill became part of an American silk throwing dynasty with 14 mills, 6,000 workers, and $50 million dollars in annual sales. In the 1930s, the company added rayon to its products. With the 1940s came wartime silk shortages and the rise of synthetic fibers. The dynasty collapsed several years before the last production run in 1957, when reelers, coners, and testers walked away from what was now General Textile Mill and never returned. The doors closed and time froze. This remains the only intact silk mill in the United States.

The raw silk came from the Orient to the West Coast of the United States via fast sailing ship, and was transferred to express trains for the journey east. This was a time-sensitive cargo, and the silk trains were given priority over even passenger trains. The manufacture of silk was a multi-step process. The fibers were reeled into skeins containing 1 to 2 ounces of silk each. The skeins were bundled into large bales of 200 pounds each. These bales were imported into the United States.

Raw silk was too coarse to be worked as it came from the bales, and it contained a natural gum that had to be removed. Thus, the first step in the processing of silk was to wash it in large vats. Next, the silk was wrung out and allowed to dry. This process was known as "throwing." From there, the silk went into the winding process. The skeins were opened, placed on an apparatus called a "swift" and attached to a spool. The silk was wound onto the spools. In some cases, depending upon its intended use, the thread had to be doubled. If so, two or more threads were united on one spool. The spools, or bobbins, of silk were then twisted, reeled and made into new skeins that were taken to be dyed.

The silk mill was run by electricity, with an extensive arrangement of overhead line shafts to drive the mills by leather belts. The throwing mill produced silk thread from the raw silk, which was then wound on bobbins. This product was in turn shipped, mostly by rail, to weaving plants near Reading, Pennsylvania to make cloth, ribbon, and finished products. The silk mill had it's own rail siding.

Afterword

We are indeed fortunate that the Savage facility has been saved. It is a standing monument to the hundreds of men, women, and children that labored there. It spans our history from before the Revolution to the current time. It reminds us of the march of technology, from water power to steam power to aps for our phone. Go there for lunch. You can just hear the sounds of the lineshafts, driven by the water wheel, and powering hundreds of cotton processing machines.

Bibliography

Baer, Mary Baldwin, Baer, John Wilbur *A History of Woodward, Baldwin & Co.,* 1977, ASIN-B0006CU0MA.

Bookwalter, H.,Wesley Leffel, J. *Leffel's Construction of Mill Dams: And Bookwalter's Millwright and Mechanic*, 2012 (reprint), Ulan Press, ASIN: B00A45IGXY.

Bullock, James G. "A Brief History of Textile Manufacturing Mills along the Jones Falls," 1970, avail: archives.ubalt.edu/bvc/pdf/7b-1-1.pdf

Catling, Harold *The Spinning Mule*, 1986 The Lancashire Library. ISBN 0-902228-61-7.

Chapelle, Howard Irving *Baltimore Clipper Its Origin and Development*, 1930, Marine Research Society, 1st ed, ASIN-B000MFCM72.

Chidester, Robert C. *A Historic Context for the Archaeology of Industrial Labor in the State of Maryland,* Maryland Historical Trust, 2004. avail: University of Maryland Library.

Etchells, W. *The Cotton Spinner's Companion,* reprinted 2016, Amazon Digital Services, ASIN-01J0050T6.

Evans, Oliver (ed), Evans, Cadwallader, *The Young Mill-Wright and Miller's Guide*, 1848, reprint 2015, ISBN- 978-1298491589.

Fairbairn, William *Treatise on Mills and Millwork*, 1863, London: Longmans, Green and Company.

Filby, Vera Ruth, and Filby, P. W. *Savage, Maryland*, 1965, for the Savage Civic Association, ASIN-B0007J3V1Y. (Available, JHU Library).

Fitton, R. S., *The Arkwrights: spinners of fortune*, 1989, Manchester University Press, ISBN-0-7190-2646-6.

Gillmer, Thomas C. *Pride of Baltimore: The Story of the Baltimore Clippers: 1800-1990*, International Marine Publishing Co. 1992, ISBN-0877423091.

Gilbert, David T. *Waterpower Mills: Factories, Machines & Floods At Harpers Ferry, West Virginia 1762-1991*, 1999, ISBN 0-9674033-0-8.

Greff, Jacquelin *Fell's Point* (MD) Images of America, 2005, Arcadia Publishing, ISBN-073851845X.

Harwood, Herbert H. *Impossible Challenge: the Baltimore and Ohio Railroad in Maryland*, Barnard Roberts & Co. 1979, ISBN- 0934118175, p.497.

Hughes, William Carter *The American Miller: and Millwright's Assistant (1855)*, 2010, Kessinger Publishing LLC, ISBN – 978-1104477752.

Jefferson, Sam *Clipper Ships and the Golden Age of Sail: Races and rivalries on the nineteenth century high seas,* 2014, Adlard Coles Pub, ISBN-1472900286.

Kennedy Alexander; Catòlica, Església *The Practical Cotton Spinner Shewing The Methods Of Calculating The Different Machines Made Use Of In A Cotton-spinning Factory: With An Accurate And ... Of Changing Systems To Any Grist Required,* 1923, reprinted 2012, Ulan Press, ASIN-B00AIJ5Z4Y.

Macaulay, David *Mill*, 1983, Houghton Mifflin, ISBN-0-395-52019-3.

Marsden, Richard *Cotton Spinning: its development, principles an practice,* 1884, George Bell and Sons, reprinted Bibliobazaar, 2009, ISBN-1113146443.

Nelson, Carl A. *Millwrights and Mechanics Guide*, 1972, 2nd ed, Audel, ISBN – 978-0672232015.

Sadler, Samuel B. *The Art and Science of Sailmaking,* 2015, Sagwan Press, ISBN-1340178303.

Scott, Robert, *The Practical Cotton Spinner, and Manufacturer: The Managers', Overlookers', and Mechanics' Companion. a Comprehensive System of Calculations of Mill ... to Which Are Added Compendious Table*, 1923, reprinted 2012, Ulan Press, ASIN-B00A44UOVM. Avail: books.google.com

Scott, Robert and Bryne, Oliver *The Practical Cotton Spinner, and Manufacturer; The Managers', Overlookers', and Mechanics' Companion*, London, 1851, reprinted 2016, Wentworth Press, ISBN-1363527967.

Smith, Merritt Roe, *Harpers Ferry Armory and the New Technology: The Challenge of Change*. Cornell University Press, 1980, ISBN-0801409845, p.286.

Stakem, Patrick H. *History of the Industrial Revolution in Western Maryland*, 2012, PRRB Publishing, ISBN-9781520215761.

Stakem, Patrick H., Sparber, Andrew "The Cumberland Cotton Factory," 2018, publication, Journal of the Alleghenies, 2019.

Steel, David R. *The Art Of Sail-Making: As Practiced In The Royal Navy And According To The Most Approved Methods In The Merchant Service* (1843), reprinted, Kessinger Publishing, LLC, 2008, ISBN-1437065732.

Wallace, Anthony F.S., *Rockdale: The Growth of an American village in the early Industrial Revolution,* 1978, Alfred A. Knopf, New York.

Winyan, Soo Hoo "At Maryland's Savage Mill, history and commerce converge," Washington Post Magazine, 04/27/2016.

Resources

National Register of Historic Places, Inventory- Nomination Form. Maryland, HO-213 (The actual record has not yet been digitized as of 2017) available: https://npgallery.nps.gov/NRHP

Bollman Suspension Truss Bridge, Maryland Historic Trust: http://mht.maryland.gov/nr/NRDetail.aspxNRID=101&COUNTY=Howard&FROM=NRCountyList.aspx.

Savage Industrial Center, Maryland Historic Trust. http://mht.maryland.gov/nr/NRDetail.aspxNRID=217&COUNTY=Howard&FROM=NRCountyList.aspx

US Patent and Trademark Office, Patent of Wendel Bollman, 1852, number US 8624A. Available, GooglePatents.

National Park Service, U. S. Department of the Interior, Salem Maritime National Historical Site, Salem, Mass. *The Great age of Duck*, Vol. VII, N. 4, September 2005.

http://www.historylearningsite.co.uk/britain-1700-to1900/industrial-revolution/the-cotton-industry-and-the-industrial-revolution/

"Growth of the Cotton Industry in America", avail: https://www.sailsinc.org/durfee/earl2.pdf

U.S. Geological Survey, National Water Information System, Little Patuxent River at Savage, MD, https:.waterdata.usgs.gov/usa/uv?01594000.

Water Wheel Factory (power calculations) http://www.waterwheelfactory.com.

U. S. Patent 84048, Annual Report of the Commissioners of Patents for the year 1911, Savage Manufacturing Company, Process for drying paper using cotton Duck.

Knowles-type planes, Mid-West Tool collectors Association, see: https://www.mwtca.org/the-gristmill/sample-articles/93-knowles-type-tools-planes-spokeshaves-and-a-scraper.html

Warfield, Joshua Dorsey *The Founders of Anne Arundel and Howard Counties, Maryland: A Genealogical and Biographical Review from Wills,*

Deeds and Church Records, 1905, reprinted, Wentworth Press, 2016, ISBN-1362551783. (543 pages) avail: Google Books, see, Savage Factory, p. 364.

The Baltimore Clippers

Bozzuto, Barbara, Gilmer, Thomas, Pease, Greg *Sailing With Pride*, C.A.Baumgartner Publisher, Baltimore, MD. 1st ed, 1990, ISBN-0962629901.

Chapelle, Howard Irving *The Baltimore Clipper*, 1930, Marine Research Society. Reprinted by Dover Maritime, 2012, ASIN-B00FBHND0G.

Gilmer, Thomas C. *Pride of Baltimore: The Story of the Baltimore Clippers, 1800-1990,* 1992, International Marine Publishing Co., ISBN-0877423091.

Pride of Baltimore-II, http://www.pride2.org/

Relevant to the Savage Manufacturing Company

https://www.revolvy.com/topic/Savage%20Mill&item_type=topic

American Textile Reporter, For the combined Textile Industries, Vol 13, 1899. avail, Google Books.com

Relevant to Woodward, Baldwin & Co.

https://trademarks.justia.com/owners/woodward-baldwin-co-inc-45824/

Maryland Historical Magazine, Fall 1992, Volume 87, Number 3.

Relevant to Whitehall Mill

Whitehall Cotton Mill,"Explore Baltimore Heritage", avail: https://explore.baltimoreheritage.org/items/show/433

Relevant to Savage Railroad

http://www.abandonedrails.com/Patuxent_Branch

http://www.trainweb.org/oldmainline/wasspur2.htm

http://www.historicalmarkerproject.com/markers/HMDI1_the-patuxent-branch-of-the-b-o-railroad_Columbia-MD.html

http://historicbridges.org/bridges/browser/?bridgebrowser=maryland/guilfordpratttruss/

http://find.mapmuse.com/details/inline-trails/338615851/patuxent-branch-trail,-md

Workers Housing, www.mht.maryland.gov, document HO-215.

Manor House, www.mht.maryland.gov, document HO-218.

http://www.cottontimes.co.uk/

Avondale Mill - https://mht.maryland.gov/nr/NRDetail.aspx?NRID=559

Wikipedia, various.

Glossary of Terms

Backwater – water that backs up into the tail race and wheel pit, when the water source is high.
Bale – a package of cotton, about 17 cubic feet, weighing about 500 pounds.
Bearing – Iron support in which a journal turns. Lined and lubricated to reduce friction.
Bevel gear – tooth bearing faces of the gears meet at angles.
Bobbin – large spool, to hold the yarn. Held by a spindle.
Boll – the seedpod of the cotton plant.
Bollman Truss bridge – pre-fabricated bridge, built at the B&O Shops in Baltimore. Invented by Wendel Bollman.
Breast – curved floor of the wheelpit, upstream and below the breast pit.
Breast wheel – a water wheel with a continuous row of buckets
Cap log – a big log at the top of a dam to minimize wear on the dam crest.
Carding – a mechanical process to dis-entangle fiber, and combine multiple fibers.
Clerestory – upper part of a building with windows to allow light to enter.
Coffer dam – temporary structure, to divert water.
Cotton – almost pure cellulose, around the seeds of a cotton plant.
Cotton duck – heavy-weight (30 oz to the square yard) cotton fabric; sailcloth.
Crown wheel – gear with teeth that are at right angles to the face.
Cupola furnace – smaller, cylindrical furnace to melt cast iron.
Drawing – producing uniform strand of fibers by passing them through a series of rollers.
Felloe – outer part of a wheel, that the spokes are attached to.
Fly-wheel governor – a feedback device to automatically regulate the speed of a water wheel, turbine, windmill, or steam engine.
Foundry – a facility for making castings, from pig iron.
Fulling - a waterwheel at the bottom of a chute, turned by falling water.
Grist mill – grinds grain into flour. Powered by wind, water, or steam.
Gudgeon - A *gudgeon* is a socket-like, cylindrical (i.e., female) fitting attached to one component to enable a pivoting or hinging connection to a second component. This carries a pintle fitting, the male counterpart to the gudgeon, enabling an inter pivoting connection that

can be easily separated. (Wikipedia)

Head – height of water; defines the amount of work that can be done.

Headrace – carries water from the source to the water wheel or turbine.

Jenny, spinning – a spinning frame with multiple spindles. James Hargreaves, 1764.

Journal – Iron shaft that turns in a bearing. Axle.

Lantern pinion – two parallel disks, with a series of equally spaced bars in between.

Line shaft – the main horizontal shaft that is turned by the power source, and distributes power.

Loom – where yarn is woven into fabric.

Mansard roof – hip roof, characterized by 4 sloping sides, becoming steeper half-way down.

Mill privilege – the ride to divert and use part of a river's flow for commercial purposes.

Mill Wright – engineer, craftsman, master mechanic

Mule spinner – a machine to manufacture various weights and strengths of yarn. Automated.

Overshoot wheel – Water introduced at the top of the wheel.

Penstock – a channel to carry water to a wheel or turbine.

Picking – process to clean raw cotton before spinning. Originally manual labor, later automated.

Pinion – usually, the smaller gear, which drives other gears in the gear train.

Pirm – weavers bobbin, or spool.

Raceway – channel for water, to and from the wheelpit.

Roving – a long narrow bundle of fiber. After carding, the fibers are more parallel.

Shuttle – carries the weft, in the weaving process.

Sizing – treating with a starch to cut down on breakage.

Slabbing – removing the outside section f a log, to squarw it.

Spillway – a channel to allow draining the head-race for maintenance of the dam or wheel. Also allows for overflow during floods.

Spindle – axle for spools and bobbins.

Spinning – process to draw and twist fibers into thread.

Tail race – carries water from the wheel or turbine.

Throstle – machine to draw, spin, and wind yarn on spools.

Trash boom – a secured log across the entrance of a raceway to catch

floating debris.
Trashrack – a filter arrangement to allow water to pass, but catch debris before it can get into the wheel pit or turbine.
Tub wheel – horizontal water wheel. Early turbine architecture.
Turbine – a rotary mechanism that produces work from the energy of fluid flow. Usually, a rotor in a casing.
Wheel pit – stone enclosure in which the water wheel turns.
Warp – threads that run the length of the fabric
Weft – threads that run across the fabric.
Wheel house – structure enclosing the water wheel.

If you enjoyed this book, you might find some other books by the author interesting. These are available from Amazon in Kindle e-book, and printed form.

Stakem, Patrick H. *The History of the Industrial Revolution in Western Maryland*, 2011, PRRB Publishing, ISBN-1520215762, ISBN-1520215762.

Stakem, Patrick H. *Eckhart Mines, The National Road, and the Eckhart Railroad*, 2011, PRRB Publishing, ISBN-1520215878.

Stakem, Patrick H. *Down the 'crick: the Georges Creek Valley of Western Maryland*, 2014, PRRB Publishing, ISBN-152021622X.

Stakem, Patrick H. *Lonaconing Residency, Iron Technology & the Railroad*, 2011, PRRB Publishing, ISBN-152028642.

Stakem, Patrick H. *T. H. Paul & J. A. Millhollland: Master Locomotive Builders of Western Maryland*, 2011, PRRB Publishing, ISBN-152019935X.

Stakem, Patrick H. *Tracks along the Ditch, Relationships between the C&O Canal and the Railroads*, 2012, PRRB Publishing, ISBN-1520216327.

Stakem, Patrick H. *Iron Manufacturing in 19th Century Western Maryland*, 2015, PRRB Publishing, ISBN-1520216246.

Stakem, Patrick H. *Ross Winans, an ingenious mechanic of Baltimore*, 2016, PRRB Publishing, ISBN-9781520207292.

Stakem, Patrick H. *The History of the Industrial Revolution in Western Maryland*, 2011, PRRB Publishing, ASIN-B004LX0JB2.

Stakem, Patrick H. *Fort Cumberland, Global War in the Appalachians: a Resource Guide*, 2012, PRRB Publishing, ASIN-B0088BWK06.

Stakem, Patrick H. *Cumberland, Then and Now*, 2011, Arcadia Publishing, ISBN-*0738586986*.

Sparber, Andrew; Stakem, Patrick H. and The *Cumberland Cotton Factory,* J. Alleghenies, 2019 publication.

For 2018 release:

Snowden's Iron Works.

Studebasker's Wagons

www.ingramcontent.com/pod-product-compliance
Lightning Source LLC
Chambersburg PA
CBHW030520220526
45464CB00006B/2874